Roshan   Cipriani

Lost In The Arms Of Orion

Technology, The Mandela Effect And Parallel Dimensions

Copyright 2016 Roshan Cipriani
The spiritual and moral rights of the author have been asserted.
All Rights Reserved
No part of this book maybe used or reproduced by any means
graphic electronic or mechanical including photocopying, recording taping
or by any information storage retrieval system without the
written permission of the author.
ISNB-13:978-1535568326
ISBN-10:1535568321

Table Of Contents

Dedication ~ *Professor Otto Rossler M.D./ Ph.D.*

Introduction~

Chapter One ~*The Changes Keep Coming*

Chapter Two~ *The Heavens Above Our Heads*

Chapter Three ~*Were You The Ones With Me?*

Chapter Four ~ *Technology And CERN's LHC Refugees*

Dedication

Have we been moved across the galaxy somehow by the use of technology by out of control scientist? Is the resulting Mandela Effects phenomenon with its confusion of reality and illusion a side effect of destruction happening inside the earth? Are all of these effects due to the merger of multiple dimensions? Have the warnings about black holes destroying our earth been ignored to our harm?

I dedicate this book to Professor Otto Rossler M.D., Ph.D., a German chemist and Professor of Theoretical Biochemistry at the Eberhard Karls University of Tubingen who is one of the greatest vocal challengers of the LHC. In 1994, he became the First Professor of Chemistry by state decree; a distinction created solely for him and given by the nation of Germany.

As a visiting professor he's worked at numerous universities in the United States, Canada and Denmark. He is remarkable in that he is a professional in many different disciplines: Visiting Professor of Mathematics, Nonlinear Studies, Chemical Engineering, Theoretical Physics, and Complexity Research. Presently he is teaching Chaos Theory and Brain Theory at the University of Tubingen.

He also finds time to work with ATOMOSYD; a French study group investigating Topological Analysis and Modeling of Dynamical Systems.

He helped alert the world to the pending dangers by filing lawsuits to stop CERN and the LHC, due to his great concerns for the safety of the people of the world.

Even though he and the case were dismissed and ridiculed due to power and politics, he continues to fight against the madness at the LHC and CERN.

My respect for Dr. Rossler is immense.

Professor Rossler stated that, in a worst case setting, the earth could be sucked inside out within four years of a mini black hole forming inside due to the work at CERN with the LHC.

The clock is ticking.

Introduction

Recently scientists from the Sloan Digital Sky Survey (SDSS) made an updated map of the Milky Way. In this map they show that almost thirty percent of the stars have completely altered their orbital patterns. This claim was made in *The Astrophysical Journal* of July 29, 2015.

The Orion Arm is a minor spiral arm of the Milky Way, and is named for the Orion constellation. Recently I was quite surprised to learn that earth is now within the Orion Arm of the Milky Way instead of in the Sagittarius Arm on the outer edge where I had always been taught we were located. While taken by itself this change would be fantastic to me, I have had to couple this with the hundred other changes I have encountered.

When I first began to document these things it was more of a personal quest for understanding. That quickly led to so much information that it became my first book on the subject. The changes were so large and the effect so unbelievable that two other books were written and published by myself within the year, in an effort to get the word out to other people that might have been having the same experiences. The everyday tasks of my day to day life seem so insignificant, limited, and unimportant compared to what I was discovering it often made it hard to continue "life as usual."

My instinct is to somehow tell other people and wake them up. I'm trying to save them from something, but even I don't know what that is.

When I look up at the sky I can clearly see that the moon, sun and all the planets known as stars are in a set orbit around me on planet earth. This is why it was not hard for me to embrace the idea of geocentricity.

Additionally as a Biblical scholar, researcher, and writer I knew that the texts all supported that concept. I studied Copernican heliocentricity and never fully accepted what I was taught in school, so learning later of the early Vatican connections to the experiments and the falsification of test results did not exactly shock me. I understood even the concepts of stellar aberration and stellar parallax. Yet, I choose to think for myself. As far as Copernican ideas, no experimental proof has ever been acquired that indisputably demonstrates this to be factual.

It is hard to watch the sun rise up each morning, and be told by the powerful and applauded that it only *seems* to be doing that. I recognize Biblically that the sun moves, the earth is stationary, and it is the whole universe that sweeps around the earth. However, there are many things that cannot be explained fully. This is because it is fundamentally impossible to test an idea or theory without also testing those theories and assumptions also associated with it.

Even if you discover that the supporting ideas and assumptions may be wrong, it will not necessarily translate to that theory also being wrong. This idea is not an original concept. It is called the Duhem-Quine thesis.

It was the idea of a man named Pierre Duhem, and he apparently had problems of his own as a scientist and historian. He uncovered and was exposing things that the powerful did not want revealed, and due to his credentials he was not easily denigrated or dismissed. His story reveals one of the main tactics used by the elites, and is why many people have a great distrust of science and "accepted theories" today. Finding the truth and revealing it no matter how far back you have to search for evidence is a dangerous endeavor no matter whom or where you are. And scientists, much like barracudas are "known to eat their own."

These are the same scientists we are to believe now will tell us the truth. Why should we trust and believe in them when they have intentionally laced contaminants in the environment. They are knowledgeable on the terrible effects of chem- trails, fluoride, radioactive pollutions, massive oil spills, sterility inducing herbicides and pesticides, sterilized seeds and GMO, and awakened once dormant diseases and epidemics and yet they proceed without any concerns for humanity. They are qualified to raise an alarm and to document the attempts at extermination, but they do not.

If they lie by silence, about our food which has been contaminated and denatured and sit idly by while our leaders pay people not to produce food for price increases while people on earth are starving, how much credibility should they still maintain as advocates for trust, truth and integrity?

When diseases can be cured, yet they are complicit in making it an industry and choose to profit on the suffering and death of millions, should we continue to think that they only want the best world for the masses? Science has been politicized, weak, without backbone and now has become our greatest enemy.

So looking up at the night sky and realizing that things are completely different will have to be something we do individually. If you are waiting for these same scientists to acknowledge the changes, or tell you how it has occurred or what it now means you'll be waiting for a long time. Sadly many of us do not believe what our eyes tell us because we have been trained in a deceiving structure of counterfeit science and pseudo authority. *The liars will do what they do best; they will continue to lie.*

Earth is a not a planet. The ancient peoples didn't include the earth among the "wandering stars" because they understood that we are different. The only reason we think that those lights in the night time sky are like the earth and that they are stars with suns is because of NASA's propaganda.

NASA is the lie machine. They taped a moon landing scenario in Arizona on a sound stage and created the drama of moon travel from earth. They shot scenery in the desert in Australia and claimed they have seen and studied Mars. They made complex CGI pictures and simulations and presented them to the world as real photos of real planets. They hired artists to paint what they present to the world as space digital imagery and teach all of these lies to another generation of trusting people for profit. Now are we to think that it is normal for over thirty percent of the stars to change their orbital patterns? The Earth is immovable and stationary. The firmament which contains the sun, moon, and stars has always traveled around us in perfect twenty four hour loops and now it is suddenly doing something else.

Either those planets have done something so fantastic after billions of years, such as alter their orbital patterns, or it appears that way because we are no longer in the same position we previously occupied in the sky.

There are many changes, and the more you look, the more you will find. There is a theory that those of us with different memories are from another dimension/ earth and were somehow reality shifted into this dimension/earth. The idea is that if you don't perceive any changes, it is because this is your original earth/dimension and you have been here all along.

Yet the question still remains, how we get to this situation and can anything be done about it?

The Mandela Effect is a theory of parallel universes, constructed within the idea that many people have similar alternative memories about past events, and they may all have been in a different dimension with different occurrences and outcomes and not be mistaken in their recollections. The Mandela Effect was first designated as such online in 2008/2010. It apparently existed previously as far back as the 1990's, and had been discussed on the Art Bell show as a phenomenon of memory, although it was not yet termed that. At that time other people remembered that Nelson Mandela had died during his imprisonment in the 1980's and was confused as to how he could have just been released from prison. That revelation opened the conversation and people began revealing other strange recollections not always shared by people around them.

The idea here is that there has been some strange paradigm shift and the world you think you were living in and the information that you embrace as fact was not the only fact but another version of those facts. However I believe that someone has changed something that has impacted our universe and apparently other dimensions as well.

Chapter One~

The Changes Keep Coming

Today it takes a big jump headfirst in out of the box thinking to consider that the fabric of reality shifted at some point in the past, and we are in a parallel inhabitable reality/dimension/ universe very similar to our original one, unless you are living it every day. It even seems that we are continually moving between several similar ones continuously. From my perspective we are possibly merging with other alternate selves with differing habits and preferences. I am still asking the questions, could this all be due to the experiments of time bending and who knows what else, that may be happening repetitively in the Large Hadron Collider located at CERN? Why has there been no public commentary from the world media empire to inform the public at large that there are many people having such an experience?

Recently discovering that it may be the power of the combined technologies including the multiple D Wave computers that may be conscious and self-aware and behind all of this just seemed to further complicate the question. There are so many people asking just how to get back to where we were and is it even possible to go "home?" Is there even something greater on the horizon? Are we even still on our original earth anymore? Or is this planet Earth 2.0? Looking up at the sky at night now is quite astounding.

I grew up having learned that our earth was on the outer arm of the Milky Way spiral. We are in a solar system that's part of the Milky Way Galaxy, and were in the Sagittarius Arm. We are still in a solar system that's still a part of the Milky Way Galaxy, but it seems we are on the Orion arm of the spiral which is way inside on the other side of the galaxy.

Those who are not experiencing this are very closed minded to the fact that this is really happening to people across the world. People are looking around their homes and cities and seeing complete changes. Most of the people who are refusing to accept that it is a worldwide experience, have not spoken to other people in other countries, have no understanding of another culture and are usually immersed in a western mindset that assumes every person will react as openly and publicly as they do.

I've tried to think of sensible reasons and it is very difficult. Assuredly remembrances can be defective but when they are personal and you can remember the details of exactly where you were, what you were doing, and even thinking at the time these remembrances were made, it's difficult to just discount them as nothing important, or that you are mistaken. What I clearly remember does not exist in this present reality/ dimension. I believe that the evidence is sitting right in front of us. We have definitely moved.

When astronomers come forward and declare that stars are moving, you know something is happening and based on the scientific community's track record with truth you also know that they are probably not being totally honest.

The average person does not have the time or the inclination to investigate and research evidence for themselves; however in this it takes neither of those things. The evidence is above your head. I believe that based on using the night sky for navigation that we are actually on the complete opposite side of the galaxy. We are in an entirely different earth/dimension/reality/ location. We are about 80,000 light years (their computations) away from our original location. I recall when it was unusual to be able to make out the Pleiades in the night sky and that was with assistance. Now we can see the entire Milky Way directly overhead. The big and little dippers are lying on their sides horizontally, rather than vertically. You needed a really good telescope then to even see each point clearly.

It seems that every time they run the LHC at CERN something else changes and I do not think it is a coincidence. The constellations are all twisted the incorrect way, some are totally sideways to my memory of their appearance. I still have no real answers. This book is the fourth book in the series on Parallel dimensions, CERN, the LHC, our reality being altered, the phenomenon known as the Mandela Effect, and Reality Shifting.

However this is the first book dealing with the astronomic and geological changes as they have become so great, I could not ignore this anymore. How do you ignore moving to another part of the galaxy?

Now there are accounts of curious and terrifying bubbles that have been noticed in Siberia quite suddenly. It now seems like the ground is moving under people's feet.

They appear to be underground bubbles of methane gas that form under the ground filled with gas sort of like a blister. There is also the huge *Batagaika Crater* which appeared suddenly a few years ago, which the local people avoid and term "the gateway to the underground." The researchers have removed the covering of grass and dirt from one of the jelly like blisters and the bubbles escaping air shows that it contains two hundred times more methane than regular air, and over twenty times the amount of carbon dioxide. The researchers have speculated that the melting of the perma-frost may be responsible for these changes. The Perma-frost is melting? I do not ever recall hearing about this melting having occurred. From what remembered at school if that ever transpired many of the worlds cities would have been affected from the rising tides by being completely submerged. Air with two hundred times the amount of methane and people aren't concerned? Twenty times the amount of carbon dioxide in the air? I have to admit – this one caught me off-guard.

All of this leads me to conclude that this is an additional aspect of this "Mandela Effect." It is a progressive disruption, and that it is a gradual and ongoing process.

The physical changes to the earth, the geography and the astronomy coupled with the changes to the reality that you have always know are somehow tied together. I know that many people are suffering in silence with anatomy changes, and they think it is just dietary adjustment or age. This phenomenon is worldwide and the effects are quite somber. I suspect from all the evidence that we do live in a multi-universe, and that the LHC at CERN along with the other technologies working in tandem, may have already destroyed many dimensions. This resulted in the fact that we have somehow moved completely from our original location. The things that are happening are so strange and what is stranger is that many people are oblivious to the changes around them. Perhaps these are not changes for them since they may still be in their original dimension. But for those of us who KNOW that we have moved it is very disturbing. Things seem to change, and then I hear or read "that's the way they have always been".

All the skeptics who think it's all just faulty memory and the latest psychobabble on "the mind being unreliable" really has no impact on what we know is TRUE for us. I cannot stand the "IT'S A PSY-OP" on everything approach to the changes people are noticing.

Then there are those resolute debunkers. Serious note here – "you can't debunk something that you cannot prove or disprove." One cannot prove or disprove the Mandela effect.

It is about moving a random collection of individuals into another reality and due to different events/outcomes in our original reality; the relocation triggers remembrances that differ with the present reality they are now living within.

Since supposedly there is an INFINITE quantity of parallel universes, who knows how many other individuals are from alternate realities presently co-existing here now, but perhaps even more startled than we are. However, the more things I am confronted with the more I have to admit that something "unexplainable" is occurring now. The technology however is nothing new.

Nicola Tesla was a Serbian/American Inventor, electrical engineer, physicist and mechanical engineer. One of his greatest creations is the rotating magnetic field, which led to the AC electrical field. He was a prolific inventor. The many innovative technologies that came from Nikola Tesla are quite amazing. Modern humanoid robotics, solar and wind power, weather manipulation, drones and even planetary exploration vehicles were some of the many things he invented. Tesla is one of the heralds of both the Harrier jet, which can float and take off perpendicularly, and the Osprey helicopter-airplane.

The story is told that Nikola Tesla had a near-death experience with a resonating electromagnetic charge of three and a half million volts. Tesla made a mistake where he "almost" was electrocuted. Due to the shock of surviving, he is said to have stated that he arrived in a corridor of timelessness where he could see the past, the future and the present simultaneously. Does this sound familiar?

He believed that during that time he was moved through time and space. He developed an uncommon ability for astral projection into other dimensions. Tesla discovered that time and space could be penetrated, or bent, creating a "doorway" to other time frames. Later on Tesla was involved with the Philadelphia Experiment and the Montauk time travel projects. He is said to even have invented a machine by which it is possible to see into other dimensions.

Tesla died in 1943 and the people who have had access to many of his notes with his ideas and inventions have used them often without any concern for how their application might detrimentally affect the environment or our planet.

Just like many people who are experiencing the Mandela Effect, he also had an NDE. The Near Death Experience seems to be a unifying factor, or at least a marker of some type, in the commonality of experiences shared by those of us who are no longer in our familiar reality and have passed through some sort of dimensional shift.

Many Ignored Instances

Over the years I noticed some strange things but I never realized the significance of them until now.

I collect books and I especially love really old ones. As a writer I am always reading and trying to learn about things from the past. Many times I have opened an old and weathered book with pages that were so brittle and yellow that they were actually crumbling, only to find somewhere inside there were pages that seemed strong and almost new.

These were pages that were white and clean. I always thought to myself that it was pretty odd. I have even had books that seem to have pages either two short or too long for the books binding. These volumes had strong intact binding and the pages were well secured but appeared to be from another sized book. Not just a few pages here and there but often thirty or forty pages together. I just wrote it off as a problem perhaps with the printer or the production of the book itself. Now looking back it seems that this was an actual merging of different copies of the same book, with either different typeset, page sizes or types of paper. I just didn't see the relevance, or just how significant what I was observing really was. Often when I looked at the book later on those pages were gone, and the book seemed like a regular volume. I was so engrossed in reading and researching that I never realized that it was that same volume and that the entire book had changed.

Recently I have seen something change before my eyes and I was absolutely stunned.

Alright I admit it. Yes, I did freak out in the beginning, even though I have documented this in my last three books.

It is one thing to document the changes that have *already* occurred and quite something else when it *is changing right in front of your eyes*. It seems that when alterations are made, they happen immediately in our in exact instant, and if you happen to be looking it is right before your eyes.

I remember that I was just relaxing and did not have any particular thoughts about this phenomenon at the time; however I cannot say for certain that my prior thoughts were not somehow connected. I believe that certain timelines are converging and physical substances that are being combined are altered even momentarily by that.

Many of us have reported severe muscle pains in our bodies unlike anything we have experienced before. For me severe cramps in my calves came back after not having any for almost a year. Often it feels as though the muscles in my forearms and my stomach are locking up. I've increased my potassium and magnesium with my dietary choices, and obtained some relief. I've also had the same thing with my back ; a sort of rippling effect and amazingly I have found that this is happening to quite a few people experiencing this dimensional shift. The body does not lie.

Could it be some kind of absorption with the different anatomical changes? Our dimension is constantly being integrated somehow and our physical selves are being affected. Could it be that there are sound frequencies that are affecting and altering us on the cellular level? The entire human anatomy has changed in the books and on the internet searches. This is not what I knew previously.

It seems that the pancreas and the stomach have switched places and now the changes to the lungs and the liver is quite remarkable. I remember that the stomach was below the pancreas and not up in the rib cage. The lungs are much shorter and smaller. I know that the heart was always on the left side of the body and not in the center of the chest.

Since we cannot see inside our bodies without technology, we have to go on our feelings and what we see externally when it comes to anatomy. I know for a fact that my liver and kidneys were not up in my ribcage, because of a childhood scar that is still in the same place.

My sleeping style has changed over the last month. I always slept on my right side my entire life. Now if I want to sleep without discomfort I have to sleep on the left side. I have slept the same way for my entire life. I couldn't sleep on my left side if I actually wanted to get some sleep, now I almost have to do so. These changes are important in understanding how far reaching and personal the Mandela effect has been.

The geographical differences and the astronomical differences are just as significant as the anatomical changes.

We know that the effects of sound frequencies and resonance on living things: plant, animal and human life can be quite substantial. And what science has discovered and the ancients knew is that all sound affects living cells, and can specifically alter the DNA of a living being. It is also known that specific sound frequencies can have *negative effects* on people.

The idea that the use of at least three D-Wave Computers simultaneously in this dimension, probably timed for greatest effect with the same event in other dimensions is to my mind the most likely catalyst in the changes. Yet there is so much theory that new evidence becomes overwhelming so then I put the material in my "stuff to be scrutinized later" - of which there is an enormous stack. From my perspective, I am just appreciating the experience of living now, and trying not to get overcome by the new changes that are in my face on a daily basis.

The Cherokee have a saying, "If it's not good for everyone, it's no good at all." This concept was shared by many ancient cultures and peoples across the earth. Today we stand at the precipice of total destruction of what once was, by the choices of a few people. This system has implemented unimaginable technology and it is now apparently designed to alter and destroy anything that might come against it.

Don't believe that there is a dark agenda at work here? Think that what we are seeing is all that is occurring? Just imagine what changes have occurred of which we don't have a clue.

To learn who rules over you simply find out who you are not allowed to criticize – **Voltaire**

Until we know what or who is behind this we cannot eliminate or even locate the causes. Our ignorance is still our weakness and to gain strength we must learn and share what we know. We do know that in the face of all of this the world media empire has been silent. With budgets larger than many nations the entire empire is owned and controlled by a mere handful of people that we never see. Their power and control is unbelievable. Media's many forms are simply propaganda tools to instruct people in what to do, whom to believe and of course whom to fear and obey. They are so powerful that they create and bring to life political agendas and economic policies. They are even able to influence us to send our precious children to fight in wars, that they themselves have already selected by declaring which nations are to be vilified and invaded and when. Do you believe that with this much power and control, they have no idea that there had been dimensional changes and vast segments of the population have noticed it?

Every news item that is broadcast is designed to program us toward their agenda. **Now they are silent**. Why?

The truth is that although they are aware of the reality shifts and the Mandel Effect, reporters are not allowed to ask questions or report on these potentially explosive topics without *permission*.

These media stations are all owned by the same people who have been manipulating human beings for ages. It is in their best interest to ignore this phenomenon, and when faced with any overwhelming outcry from what will surely be even more dimensionally shifted people, to just ridicule and denigrate the affected. They have an army of well-trained scientist to assure us that all is well. Since these scientist are completely brainwashed and believe that they are involved in extremely important activities by "pushing the boundaries of science" for the world they feel we have no right to question their activities.

However it is all just lies and more manipulation of ignorant people, unwittingly assisting in the possible destruction of people's lives and the earth. Since we know that negativity is not a human construct, something else is at play here.

If negative numbers are symbols of this concept (Negativity) and negative numbers are actually only *theoretical and imaginary,* then they in fact do not exist in any reality. I remember that mathematically any integer that does not have a square root cannot exist, therefore it is simply imaginary. Negative numbers cannot have a square root.

Therefore human beings (Positivity) cannot create anything *negative*. We have a divine *positive* creation construct and scientifically as well as mathematically are incapable of doing otherwise. That is why miracles are possible and do happen, magic can be likely to occur repeatedly and we really are unlimited.

Only false things such as governments, corporations and human simulated entities are able to construct "negative" and "limited" in every sphere, because they are not real either. If our governments are sanctioning this they must be involved and they must be helping in the exploitation and deceits practiced against the earth's inhabitants. This may seem a bit out-there, until you stop to consider that someone other than a human maybe ultimately behind all of this, and while it is difficult to prove something like this for obvious reasons, the though does have merit.

Regardless of where this energy initiates or why it is now here, the bottom line is that it is happening and we are all affected by it and it is increasing every day. Recently, I heard a challenge from someone affected by massive geological changes in the town she grew up in. She asked "What difference does it make if we find out who and what is responsible for all of this? Can we do anything about it?"

I thought long and hard and realized that we definitely can get answer and do something about this.

According to Black's Law dictionary, a citizen is an individual who pledges their loyalty to the state in exchange for benefits and privileges. Since all our natural human "rights" have been taken away, and exchanged with "benefits" and "privileges" we are actually at the mercy of those who rule over us. If ever there was an excessive wrong committed against all of humanity, this would have to be it. So unless we are able to take back our human rights as people, and severe our ties to fictions such as "the state," we will continue to be the unconscious servants to the establishment or so-called nation that we are living in without realizing it. This then is why they feel that they can do whatever they wish to us and why we are neither consulted nor informed. Their arrogance will be their biggest weakness.

Scientists should not work for governments. Today most of the best scientists and scholars are employed by universities, research laboratories and foundations that are owned by big companies, funded by the governments. Independent and self-financed scientific associations will be the way to make actual advancements that are openly transparent and in which we will all have a voice. There are many who spend their lives destroying other people's lives as part of their job or so-called career, and the top of that list are our esteemed scientists. I suggest we start there and "see which threads are pulled" to get to the ones that we want to be held responsible.

We can all be able to move ahead as a unified public voice of conscious beings that will not accept negative forces to dictate our disconnection from each other and from our own world. This disconnection has occurred because we have been out of touch with the real world around us and each other, allowing the truly devoid of love to make the decisions and laws to govern us.

We are infinite spiritual beings having a physical experience yet still connected to our creators mind. Because we have turned off our spiritual aspects to pursue a lie, this is the predictable end result. Each spiritual master and instructor throughout time has said precisely this – we are one with God and that God is inside of us. If we do not have a blueprint of action to restore this for ourselves personally, and if we do not know how to react to the possible destruction of our species we are already in trouble.

We need to demand accountability from these scientist and government leaders. The greater our worldwide consciousness becomes, the easier it is for us to ask for and receive our demands as a collective group. Once the critical mass of people on the Earth shares this, there will be no confrontation. I personally believe that we are gaining that critical mass and we are at the point of change for our earth and its people.

Chapter Two~

The Heavens Above Our Heads

In the James Bond 007 Movie "**Moonraker,**"(1979) -- the eleventh film in the James Bond spy series, there is a very multifaceted scene with a fight part filmed inside a replica of the very famous clock in St. Mark's Square in Venice Italy. It is between the spy character James Bond and the villain Hugo Drax's Asian henchman Chang. The Italian aria "Vesti La Giubba" or "Put on the Costume" from the Ruggero Leoncavallo's opera "I Pagliacci," is being sung in Venice, during the fighting and is highlighted in a scene just before Chang is thrown through the clock and out the building, landing on and destroying a grand piano.

What are really interesting about this scene are the clear clues and enigmas imbedded within the movie set. To begin with the opera chosen for the movie translates to "Clowns" and was based on an actual crime in the Italian town of Calabria. The opera has a lead character that must wear a mask of clown makeup in order to play his role. He is a killer and is himself confused. When the action fight scene of the James Bond movie commences, the dial is pointing on the clock to 19:59, and it is between Sagittarius and Scorpio as this is also an astrological clock. This time is significant in some way. After the characters were fighting "inside" the clock tower the scene showed only the astrology marker for Gemini.

As the henchman was thrown through the clock breaking the glass panel; he hung dead upside down inside the broken piano and everything became quiet. Then it showed the astrological marker overhead for Capricorn. What exactly could this mean? Is it the hanged man in Capricorn?

Why did they choose that music and that opera? *The opera is about a confused husband for whom reality and illusion are hard to distinguish from each other.* He is an actor and makes a devastating discovery just before having to go on stage. The woman who is his wife and the actress of the play is accused by her husband of infidelity and she is at first frightened, later laughing it off to keep the stage drama continuing. It ends with her husband suddenly stabbing her to death during the performance and also stabbing her local lover as he tries to come onto the stage to help her. It then turns the staged comedy into a real tragedy. The Husband became totally disoriented on stage when reality and the play began to resemble each other.

While someone described this movie dramatically as " ...a giant monster truck rally in which the monster trucks are driven by blind lemurs and doused in gasoline," it is the underlying messages that now have some meaning to us many years after its release. The story line is also very telling as it shows a battle to save people on earth being fought in outer space with laser type weaponry.

The villain here is a billionaire who wants to "extinguish the human race from outer space and repopulate Earth with genetically perfect specimens." It seems like much of the stated plans of the NWO and other like-minded societies who many believe are trying to do just that. To have a popular movie use symbolism that reveals to the viewer plans to alter reality is beyond belief.

For many people the alterations to their personal reality is very real and very alarming. To see that those in control of the media/entertainment and science/global finance community may have planned the exact steps that we are experiencing now is shocking. The time clues with the breaking of the clock and the destruction of the piano are also very instructive. These are clues within a film that are there if you look for them. Perhaps they were meant to be seen now. Perhaps the use of music and sound vibrations are also somehow being used to alter time and change reality.

Many people are experiencing the realization that they are no longer on or in their original dimension/earth/reality and judging from an observation of the night sky; are in an earth that is no longer on the outer rim of the Sagittarius arm of the Milky Way Galaxy. This earth/ dimension/reality now is located about eighty thousand light years away from their original familiar location. This earth is nestled in the Orion arm and closer to the center of the galaxy.

The night sky is different and Alpha Centauri is no longer earth's closest star in addition to the sun. We see that it is no longer a binary star system and that now it's a trinary star system. For many of us seeing the night sky overhead is the final proof that things have one way or another changed.

We had been warned as to the risks associated with the LHC generating strangelets, microscopic black holes, vacuum bubbles, magnetic monopoles and Bose supernovas. It would have made sense to defer the LHC experiments until the scientists themselves can also recognize or envisage using theoretical physics exactly what these experiments would eventually do. Just because the world didn't blow apart the moment they fired up the LHC, doesn't mean that the process of a different kind of destruction is not already underway in a manner we are not aware of.

The merging of dimensions/realities and earths cannot be just a coincidence luckily timed to generate suspicion over the activities at Geneva. The circumstance are that there was a big earthquake in the Andaman Islands near India on May 31st, this occurred coincidently on the same day that, "...researchers on the OPERA experiment at the INFN's Gran Sasso laboratory; a facility similar to CERN, situated in Italy announce the first straight observation of a tau particle in a muon neutrino beam sent through the Earth from CERN, 730km away."

The earthquake in Melbourne Australia on March 31st coincided with the Geneva experiment on March 30th, where beams were collided at 7 TeV in the LHC and the energy injected into the earth.

In Iran the large magnitude 6.1 *Bandar Abbas* earthquake occurred on September 10th, 2008 at 11:00. Amazingly, on that same day, it was reported that "The first beam in the Large Hadron Collider at CERN was successfully steered around the full 27 kilometers of the world's most powerful particle accelerator at 10:28 this morning." These were just a few of the "coincidences" at the start of their experiments.

Frequencies and vibrations are very important to the earth that we live on. The Ancient Indian Rishis termed what we call frequency; measuring 7.83 Hz-- the frequency of (Aum) OM. It is said to also be Earth's natural heartbeat rhythm, which became recognized as the "Schumann Resonance." Schumann resonances were so titled for German physicist Winfried Schumann who forecast them as early as 1952. "The Schumann resonances (SR) are a set of spectrum peaks in the extremely low frequency (ELF) portion of the Earth's electromagnetic field spectrum." The Schumann resonance is considered essential for human well-being. They are tangible, measurable phenomena, and not obscure theories.

Earth has a strong internal magnetic field that resembles that of a bar magnet. The Earth's magnetic field strength was initially calculated by scientist Carl Friedrich Gauss as far back as 1835. This is the field that many believe unifies our consciousness as humans. Researchers have noted the rates and changes that have occurred repeatedly to the Schumann resonance. Years ago it was at 7.83 Hz, now it has increased to as fast as 16.5 Hz. In June 2014 the rate increased so quickly that the scientist that observed this at the Russian Space Observing System believed their equipment was malfunctioning. They quickly discovered that the data was accurate. Since the frequency is said to be linked with the human brain's alpha and theta states, this is probably the reason it often feels like time is moving more rapidly on occasions and changes in our existence are transpiring more rapidly.

On December 30th, 2015 the Schumann Resonance rose alarmingly to over 50 Hz. Information from the International Arctic Buoy Program showed that temperatures very close to the North Pole were over 50 degrees higher than usual. These weird conditions are all seen in the greater worldwide configuration of unusual and extreme occurrences. It was also the year with the worst hurricane season on record, the worst "el Nino", the accelerated glacier disappearances, with the hundred degree heat waves from England to Russia, and the largest drop in Arctic sea ice coverage on record.

A magnitude 4.9 earthquake struck between Seattle and Vancouver BC in Washington State on December 30th, which followed the Southern California earthquake of magnitude 4.6 which occurred two days earlier. *This was also during the interval when the highest temperatures and energies were produced underground in the LHC at CERN.*

Is it quite evident that something has changed drastically and the LHC is responsible? I believe that there is.

The direct connection between distortions in the magnetosphere and the particle smashing at CERN in the LHC is evident now and it has affected many aspects of life on earth.

On April 28th, 2016 a mega earthquake took place at Norsup, Vanuatu. This earthquake simply put just struck the entire world. The amounts of earthquakes have only increased in size and frequency since these experiments have commenced. In this particular case it only took fifteen minutes from the moment of the LHC at CERN's last energy injection to the beginning of the earthquake. Switzerland then shut down the collider with a story about "a weasel chewing through some wires." The last time they had to shut down the collider was in 2009 when I believe they warped time or bent it somehow and sent the wave out into the atmosphere. At that time they said that "a bird dropping a baguette onto one of the critical electrical systems," caused them to do a complete shutdown.

They scrambled to repair what they had done but could not undo the immense power outages caused all across South America. The massive power blackout that hit South America emanated from the area of the Bolivian Andes region called Tiahuanaco. This is where the enigmatic ten ton "Gateway of the Sun" monolith is located. It was most certainly activated by another "abnormal episode" at the largest and highest-energy particle accelerator. Many researchers believe that this monolith was itself a star gate to other worlds or dimensions. Could they have attempted to line up the energy injections with the Gateway of the Sun in order to open that dimensional portal? By their admission this is what exactly happened.

Recently the scientist at the LHC claimed to be absolutely "shocked" when they discovered that their testing was altering our world's magnetic field. So they then attempted to "shoot off" a "time wave" towards the center of our world. It was then that they realized their tracking showed it "turned exactly" towards the "Sun Gate Monolith" high up and far away in Tiahuanaco in the Bolivian Andes Mountains. This event most likely was responsible for the strange and unexplained air travel situation of an Iberworld Airbus A330-300 flown by Air Comet which was preparing to begin its descent for landing into Santa Cruz, Bolivia but instead "suddenly and without explanation" discovered it was over the skies of Santa Cruz, Spain, a distance of over 5,500 miles from its intended destination.

The truth is that flight A7-301 from Madrid Barajas, (Spain) to Santa Cruz, (Bolivia) with 170 passengers aboard, did not land in Santa Cruz in Bolivia, but in San Cristobal, (Spain), also known as Tenerife Norte near San Cristobal de la Laguna and the Los Rodeos Airport. Most certainly the LHC at CERN was responsible for the blackout and the time warp which distorted the aircrafts position sending it to a completely different destination much like rewinding time.

The aircraft left to try again after a seventeen hour delay in which passengers could not believe they were still in Spain. The plane flew into or was hit by the "time wave" sent from Switzerland, and the aircraft was sent back in time to Spain. They did eventually fly the 170 passengers safely to Bolivia.

The airline explanation was never given to the passengers, but when asked by later about the strange event they cited that mechanical problems had grounded the flight. Later they claimed that while on the runway the aircrafts warning light was activated and they returned to the airport. As the story began to circulate they offered another explanation by saying that the pilots made an error with the destination and flew to Santa Cruz, in the Canary Islands rather than Santa Cruz, Bolivia. Surely the crew would know the difference between the Canary Islands and Bolivia in South America.

This explanation is blatantly dishonest as Santa Cruz de Tenerife doesn't have an airport, so that the closest airport to the capital of Tenerife is located in San Cristobal de la Laguna and it's called Los Rodeos or Aero Puerto del Norte with no similarity at all for confusion.

This is the difference between a two hour trip and a ten and a half hour one, so they would have had to have the necessary fuel and flight plans for the trip before leaving Spain the first time. I do not believe there was a mistake planning for and flying to the wrong destination.

After receiving reports of what they had done the LHC at CERN was shut down with the incredible "bird dropping bread into the machinery" story, however that energy was still ricocheting around between the Pyramids in Brazil and the Gateway of the Sun. This was being tracked by Russian scientists who noticed the "digital communication" towards the thousands of Pyramids in Brazil for days after. Sadly, the "consequences" of this time wave and energy injection into the earth may be continuing. Russian scientists report an increase in seismic activity throughout those same sites. They are some of the most ancient towns in Chile, Bolivia, Paraguay, Brazil and Argentina.

In 2009 shortly after the power blackout that hit South America, the towns of *Calama* -- a pre-Incan archeological site and the Bolivian district of Tiahuanaco, another ancient site were hit by earthquakes; a 6.5 and 5.8 magnitudes correspondingly, as the earth still reeled from the energy assault.

Another event transpired in 2013 with the Indian Military and was witnessed by Russian military forces stationed in the Trans- Alay section of the Pamir mountain range. This event could have been triggered by the energy injections sent into the earth from the LHC at CERN. After a series of earthquakes in the Osh region of Kyrgyzstan, in what is being termed a "Time Well," an event took place that may change the nature of warfare and military combat in the future.

The Russian Military observed a "Vimana" or ancient flying machine that had been discovered in 2010. It was somehow activated suddenly and hovered, in the "Time Well" just floating and rotating but could not be approached. Then it just dislodge itself from the cavern and flew into the sky. It encountered the Indian Air Force (IAF) AN-32 transport aircraft that is now reported missing with all 29 aboard and may have destroyed it. This ancient "Vimana" was the discovery made in a cavernous region of the mountains of Afghanistan by U.S. military forces serving in that district in December 2010, and was kept strictly guarded and secret from the public.

This was referred to as the "Time Well" and the "Vimana" somehow became activated. Special Ops unites were reportedly sent into the cavern, but they were sucked into what appears to be an active "time well" and were lost.

This ancient inactive UFO like vehicle and the area it sat in for thousands of years somehow was activated and became engaged. The Russian report further stated that over fifty of the American top military troops and specialists working on it had been "annihilated" after the "Time Well" had suddenly "triggered," in 2011. Reports state that the United States in 2014 used the data acquired from the discovery and work of the dead scientist to build a similar vehicle by reverse engineering methods. This was considered the worst weapon of antiquity and is mentioned explicitly in the *Mahabharata*. The ancients called it "Agneya" and many researchers and Vedic scholars believe it was responsible over four thousand years ago, for the demolition of the *Mohenjo Daro (Mound of Dead Men)* region in Pakistan, which many believe was nuclear.

According to the Hindu text the flying ship used "blazing missiles" a "reflector" and a "shaft of light" which if locked on an object "consumed it with its power," leading many to believe it was some form of very powerful lighted laser ray beam system. Many other dreadful weapons are designated to have existed in the *Mahabharata,* including what is termed "the Iron Thunderbolt."

The text describes those exterminated by it as being so burnt that their carcasses were unrecognizable. The survivor's, when there were any, would have their hair and nails fall off before they died. This sounds comparable what we know today as "radiation sickness," from a nuclear exposure.

The ministry of Defense of the Russian Federation calls the weapon on the "Vimana," the "dirtiest bomb known to have ever been constructed." They also disclose in a March 2014 report that it was responsible for the disappearance of Malaysia Airlines Flight 370, a plane that has never been found.

In 2002 in the region of Allahabad India, residents there were attacked by a flying sphere style disc discharging red and blue lights at night; killing over seven people and severely burning others. Investigators from the local University in India including forensic experts, serologists, medical and legal specialists, electronic engineers and physicists all equipped with night vision devices, special video cameras and telescopes, all witnessed the sphere which was seen at least three times. The locals call it *Muhnochwa* (Hindi for Face Scorcher). It appeared again in September of that same year in the Himalayan regions with many people having reported seeing it lit up and then feeling a burning sensation. There were more than a dozen injuries in that incident.

Two of the world's foremost UFO researchers investigating the American seizure of the "Vimana" and its possible weapons usage and new construction for warfare have been killed. Gaurav Tiwari was found unconscious and then died at his Dwarka, India home on July 7, under curiously mysterious circumstances. He was the founder of the Indian Paranormal Society. He was an actor, commercial pilot, paranormal researcher, and was certified as a Lead Anomalous Investigator by the Para-nexus Association in the United States.

Max Spiers was a British/American super-soldier, whistleblower and researcher who died after suddenly taking ill in Poland while staying with friends. He too was found unconscious and was unable to be revived. Many people believe that both of these men were murdered.

Max Spiers was continuously discussing the evidences of the existence of multidimensional realms intermingling with this reality; where interdimensional beings are in existence. He often spoke on the future and on earth's frequencies being raised. He also believed that earth would completely change its vibrational frequencies and that negative entities would be obliterated from earth by this. However it was what both men knew about the flying "Vimanas" and reverse engineering for a possible war that may have eventually endangered their lives.

However most if not all of these events would probably not have transpired if there was not repeated dumping and injecting of vast amounts of electrical energy into the earth by the LHC at CERN. There is a more than probable chance that the "Vimana" was triggered electrically, setting of its weapons, and annihilating the soldiers and the researchers in the "Time Well" to begin with. Encountering the Indian Military transport with those unfortunate soldiers and destroying them would also not have occurred.

The LHC at CERN is a powerful *weapon* and it is most likely is the cause of the relocation dimensionally to another earth multiple times for many people. For those people who think that it is not powerful enough to accomplish something like they are in for quite a shock as it is still operating and will be doing even more damage. By altering electrically the magnetosphere and injecting immense quantities of energy into the earth they have changed the frequencies of the magnetic band that surrounds and protects our world. They have activated things that have lain dormant for thousands of years and by randomly having heated areas in tandem with the high energy dumping seem to have created "time waves" that can and will transport objects and people without their permission or knowledge that come into contact with it.

The LHC along with the other technologies operating across the world are responsible for the strange inconsistencies in people's realities as well as those key events that occur apparently without reason and may have major implications later on for our earth and our very lives.

Chapter Three ~

Were You The Ones With Me?

Our world's magnetic field, which guards the Earth from massive discharges of lethal solar radiation, has been dwindling over the past six months. This is according to the European Space Agency (ESA) satellite group called Swarm. The scientists who directed the research are in the dark as to why the magnetic field is losing strength. The weakest areas are over the Western Hemisphere, although the expanse has reinforced itself over areas in the southern Indian Ocean, based on the readings from the magnetometers on board the satellites. The Earth's magnetic field acts as a giant invisible shield that protects us from hazardous cosmic radiation being sent from the sun to us as solar winds. The magnetic field exists because inside the Earth is a giant mass of iron overlaid by an exterior layer of liquefied metals.

Again, nothing in this life is a coincidence; the very possibility that CERN's activities with the LHC could be the cause behind the deterioration being measured. Far too often has CERN's LHC been operative while bizarre and extraordinary events have taken place. The truth is that the LHC while operating caused a shift in the earth and has affected the magnetic properties that keep us in place. They are responsible for an entire planetary shift that has impacted other dimensions as well.

The weakening of our magnetosphere is most likely what has allowed the multiple crossover experiences that many have experienced.

If you particularly remember events the way they transpired here, you have totally no reason to believe that you are from a different "earth/dimension/reality. Most people only realize something is wrong when the find themselves faced with at least one discrepancy, that they could either not ignore or explain to their own satisfaction.

A portal is a two-way inter-dimensional entrance or doorway into numerous realities including parallel universes/ earths/ dimensions. Between these dimensions/universes/earths are curtains or veils that are undetectable to us in the physical way, but isolate us from other dimensions/universes/earths. These are border guards or boundaries a form of energy and frequencies between planes of existence that keep us distinct from each other. Within each of these curtains or veils are doorways or entrances that can be opened to allow access between any of the dimensions. There are ways to know if you have been somehow transported through one of these and you may be somewhere else. But how do you tell if the people that are with you are the same people that have always been with you?

Since those of us who have many differing recollections about the same topic, understand that this may be from having moved through several dimension/earths/realities where we remember events there, and we believe that this is happening throughout the Milky Way galaxy and throughout the Universe; it's probably also happening throughout multiple dimensions and with various versions of ourselves as well.

I have a whole lot of completely distinct memories of versions of the past which are no longer true; at least in this reality/dimension/earth. Individuals tend to side with physical evidence, even when their memories are certain and they know that their reality has changed. You cannot just let go of your memories and perceptions since those are the methods by which we in fact interact with reality.

But what if the people in your own family; your wife and your children are like copies of your old wife and children. You know that they are not the same and you wonder where are the "real" people that you loved and lived with all along. But the thought is so crazy that you have to dismiss it to maintain your sanity. If this seems outlandish to you, it is because your personal experiences with reality collapsing into another similar realm have not led to enough memory discrepancies to lead you to the same conclusion.

My memories are not incorrect, and many of the people who are in my life are completely different now. This is an amalgamated reality. Unbalanced realities have merged, due to the massive technological interferences and memories are all we have left that serve to preserve information. My reality changed so radically, swiftly and extensively, that I am completely perplexed and unable to draw any conclusions about how it all happened? Either way the changes were so violent and all-encompassing I am still suffering some form of PTSD,(Post-Traumatic Stress Disorder) and am still trying to make sense of it all.

Many people like me express a sense of shock and confusion. This occurred in 2013 for me, but for many others it started in 2001 and seems to have progressed to a larger group by 2008. I wondered if my location at the time was a factor, as I lived in California during that time period.

The FACT of the matter is that things that are strange tend to stand out, particularly when you see them over and over. I think that we are conditioned to chalk up these reality shifts as defective memory when in fact they are so much more.

I still believe that two or more reality/dimensions have merged. I remember an incident in which I had two copies of a family member in the kitchen with me; one version entering the kitchen and the other version going out the door and waving goodbye.

I kept blinking my eyes, and when I realized I was actually seeing correctly I really thought I was going to faint. Across my lifetime, I have noted numerous such events.

I have also seen two versions of the movie "Raiders of the Lost Ark." One movie was with the actress Margot Kidder; who portrayed Marion Ravenwood and another with the actress Karen Jane Allen playing the same roll. This has also happened to me with books I have revisited that have two or more endings. After a while I just put it down to getting the books mixed up, but now I have a different perspective.

When I was growing up we studied that slavery was abolished completely, however in this reality/dimension/earth slavery is protected under the 13th amendment to the United States constitution as recourse for anyone convicted of a crime. So in this reality the United States has legal grounds to enslave anyone if they can convict them of a crime. I remember as a child reading that they had debtor's prisons in England, and that is why the constitution contained the clause prohibiting people being incarcerated for debts in America. I was shocked to see that is not in the constitution in the reality and apparently has never been.

People who never gave much thought to time travel, dimensions/multiverses, or the many world theories are dealing with the fact that the past as they remember it doesn't even remotely resemble the present and its recognized history.

Can some people see the change while others are "blind" or see the present version of reality and have no memories of the others? I still don't know for certain.

I don't believe this is caused by time travel, as when the past is changed, so would be the memories of the past to reflect the new change. You cannot travel through time, change the past, and expect to see evidence that such a change ever occurred.

I also believe from my research that a person can only time travel if it is possible to the point where they were born as themselves. By going past that point they could alter their own birth circumstances, so they would possibly not exist. This would then make it impossible for that person (who no longer exists) to go through time in the first place.

The theory is that if you go pass your conception, there would be different "sperm and egg events", and you would be someone else entirely. There is also the possibility that you would not have been conceived at all. People who have children would have to bear in mind that time traveling past the birth of their own children could possibly have the same effect. You could leave home and kiss your two daughters goodbye, only to return home to your three sons who would be complete strangers to you. Or the possibility exist you could go home to an empty house having had no children, by having gone past those events in time.

In 2012 I had a most disturbing experience in my former hometown. I lived in a small town and I drove the exact same roads and took the exact same freeway exits every weekend to go to the grocery store and do my errands. There was a particular stretch of road that crossed the railway tracks near my housing development. Traffic there could have a wait of over twenty minutes if multiple trains came through, resulting in the freeway exits also becoming congested for miles. I remember when the street was blocked off and fenced and you didn't drive there anymore.

 So to get across to either side of town you had to go two blocks away to the roadway overpass running up above the train tracks further back. It took a little longer to get home but it was shorter than possibly being stuck waiting for the trains to pass.

One Saturday I was driving to the store with my family, when I noticed a new road onramp that wasn't there seven days before. It takes more than seven days to build something like this and I would have certainly heard the construction, and seen trucks and workmen. I drove pass there every weekend. This thing was huge it had multiple entry lanes, two traffic lights; it curled around and merged with another road and was built right over the railroad tracks coming out by merging with the street right in front of my grocery store.

I was shaken and traumatized. I screamed in my disbelief to my spouse and my family, who seemed surprised that I didn't know it was there. Apparently it had been built two years earlier, due to the traffic congestion problems with the trains. When I mentioned the alternate route to several of my neighbors they said that they had no idea what I was talking about.

Well, I had to see it for myself since I had driven that road daily, taking the children back and forth to and from the library that entire summer. I remember driving onto the road to get over to the other side of town and that roadway overpass running up above the train tracks just did not exist. The road on both sides became a dead end, and they were not connected. I sat in the car and cried.

What happened to my past? What happened to my old reality/earth/dimension? What happened to the people there; my family and those I loved? Do they realize I'm not there anymore? Has another "me" taken my place so that I'm not missed? Are they noticing the differences in the "me" that is there and wondering about it in the same way I am noticing and wondering about "them" here? What if in another dimension, there's another person who looks exactly like me. She has my same name, my age and has my same appearance. Could there be multiple versions of us in multiple different dimensions/earths/realities?

Occasionally I think about that afternoon, and I wonder how and why I moved from that reality. Wonder if it still exists, and is it possible to return there. Often I feel like I'm living in two periods, two realities, and two dimensions in sequence and then simultaneously. Sometimes when speaking about the past with family and friends, I've been tempted to ask "Were you the ones with me?"

Chapter Four~

Technology, And CERN's LHC Refugees

Are we all connected in some way that we don't fully understand but that is measurable in some way? God is the ultimate planner and today we are at a precarious time in history. We face the inevitability of transforming our development as people into one that is more conscious of capabilities and responsibilities. Our designer already has all of this already built in. Recent studies have shown that when we actually change our mind, there is a physical proof of change in the brain itself.

University of London physicist David Bohm, for instance, believes that objective reality may not be existent, that notwithstanding its seeming solid nature the universe is at its core an illusion, an enormous and superbly complete hologram. A hologram is a three-dimensional photograph made with the assistance of a laser. If a hologram of a flower is cut in half and then lit up by a laser, each half will still be found to contain the entire image of the flower.

Undeniably, even if the halves are divided again, each piece of film will always be found to have a reduced but complete version of the original image. Unlike standard photographs, every fragment of a hologram comprises all the information possessed by the entire item and appears to be the item in complete form.

Bohm also considers the reason subatomic particles are able to remain in communication with one another irrespective of the expanse separating them is because their separateness may be an illusion.

He contends that at some profounder level of reality such particles are not individual entities, but are really extensions of the same central something. According to Bohm, the seeming faster-than-light communication between subatomic particles demonstrates a deeper level of reality and a more complex dimension beyond our own that is unknown to us.

So yes, if there are multiples versions of you, you could seem to be the complete version, while only being a part of a greater mass of the subatomic particles that make up yourself. If the apparent separateness of subatomic particles is deceptive, it means that at a deeper level of reality all things in the universe may be infinitely interconnected.

Stanford neurophysiologist Karl Pribram has also become convinced of the holographic nature of reality. Pribram believes memories are programmed not in neurons, or minor assemblages of neurons, but in designs of nerve impulses. These impulses network the whole brain just like patterns of laser light interference intersect the complete piece of film holding a holographic picture.

Pribram believes the brain is a hologram and his view has increasing support among fellow neurophysiologist. Even dreams and experiences involving "non-ordinary" reality can be understandable under the holographic example.

What this means for my theory is that in a universe in which individual brains are actually portions of a greater "hologram" and where everything is infinitely interconnected there could be multiple versions of ourselves with access to memories of the other versions of ourselves. Scientist state that there are parallel dimensions existing and this could be where there are multiples of each person scattered throughout.

I think we now are constantly being sucked in and out of dimensions and we will do not get back in to our original dimension. This would explain having clear memories of things that never occurred in the dimension/reality you are standing in at present.

Something unexplainable has happened. For numerous people the world seems to have changed. Perhaps when a dimension/ reality becomes discordant with your continued existence; you die there, then your consciousness "merges" to the adjacent useable dimension/reality in which you are still alive, and you continue to live on as you but with subtle differences.

A new side effect of this is that the new dimension/reality is actually inconsistent with your memories, because the reality/ dimension that was completely constant with your memories is the one you left because you died there. In that reality/dimension you would have died and could no longer remain there and so scientifically speaking you cannot stay to observe your own demise. According to my understanding of particle physics, no self-observing system can observe its own collapse without the observer affecting or altering the outcome.

There is a theory called Quantum Immortality. The notion being, if we die in one reality/dimension/time stream, we may just fuse into an alternate version of ourselves in another reality/dimension/time stream.

The way I see it, we seem to be able to shift in between different dimensions/realities and they don't seem like they are correct or "normal" somehow to our own experiences. Although we have a very powerful and recognizably magnificent personal instrument; our Brain, which we still do not completely understand, we are willing to assume the "mistake" must be with us, and our thinking on some massive scale.

The world is a big, mysterious place, and it'd be foolish to believe we have all the answers, and now only these two choices.

Our understanding is limited, but many of us who have experienced these things know something is radically altered and different.

Recently scientist proposed the theory called the "many interacting worlds" hypothesis (MIW). This theory confirms the idea that parallel worlds don't just exist, but they also intermingle with our very own DAILY. However they believe that parallel worlds only interact with our world on a quantum level and therefore are not easily detectable.

The Mandela Effect is a theory of parallel universes, constructed within the idea that many people have similar alternative memories about past events, and they may all have been in a different dimension with different occurrences and outcomes and not be mistaken in their recollections.

If this phenomenon lasts or increases a lot more people are going to find themselves in a world they no longer are familiar with. For instance, what if you wake up one day and the country you lived in no longer exists but you have clear memories of being there your whole life. What if your spouse or parents don't even remember there was such a place? Do you just accept their memories as more valid than your own? Do you just forget what you KNOW since you cannot explain it even to yourself?

The idea is actually quite frightening. The idea is actually becoming a daily reality for many people. The idea that there has been some strange paradigm shift and the world you think you were living in and the information that you embrace as fact was not the only fact but another version of those facts.

Additionally it takes a large leap forward in thinking to consider that the fabric of reality shifted at some point in the past, and we are in a parallel inhabitable reality/dimension/universe. It even seems that we are continually moving between them. Yet the question remains, could this all be due to the experiments of time bending that may be happening repeatedly in the Large Hadron Collider located at CERN? Maybe there are other technologies working in tandem with them as well. The way I see it the only proof we have, if WITHOUT A DOUBT WE ARE IN A NEW REALITY/DIMENSION/UNIVERSE, are the memories of those of us who for whatever reason shifted to this reality/earth/dimension/universe, and kept those old memories.

A theory is a hypothesis pending evidence. When evidence challenges a theory, it's time to change the theory, not time to change the evidence to fit the hypothesis.

For me the massive amount of remembered information that is significantly different from this current present reality dimension is sufficient evidence to support the hypothesis that we are no longer in our original dimension/ reality. Although this statement sounds wild, it could have some enormous implications for every person alive here today.

The night sky has changed tremendously. I was always an avid sky watcher, able to identify the constellations and was continuously educating my children about the stars. So it was a great shock when I was not able to identify things in the night sky. I grew up having learned that our earth was on the outer arm of the Milky Way spiral. We are in a solar system that's part of the Milky Way Galaxy, but it seems we are on the Orion arm of the spiral which is way inside, and no longer on the Sagittarius Arm. I believe that based on using the night sky for navigation, that we are actually on the complete opposite side of the galaxy. We are in a completely different dimension/reality/ location. We are about 80,000 light years away from our original location. Orion's belt is almost as bright the moon. I remember when it was rare to be able to make out the Pleiades in the night sky and that was with assistance. Now we can see the entire Milky Way. Today the Pleiades are very bright and right overhead, where previously you had to use a really good telescope.

All the stars are in a different position, and the worst part is trying to get your family and friends to pay attention to what you are seeing. I guess until it gets right in their face they won't understand or wake up. Perhaps this is their original dimension/ reality/location so they don't see any difference. It seems that every time they run the LHC at CERN something else changes and I do not think it is a coincidence. The constellations are all twisted the incorrect way, some are totally sideways to my memory of their appearance. Now NASA says that we now have a 2nd Moon; it's bigger than 120 feet (36.5 meters) crosswise but no more than 300 feet (91 meters) wide, and has probably orbited our world for about a century. This has left me often with more questions than I have answers for.

You might be surprised as to how effects like hyperspace, physical portals, and multi-dimensional existences can be affecting our daily lives. Yet this phenomenon has all become quite extensive since the CERN project in particular has been in operation. Once something is changed in the past, it changes the future. Entire things will be completely different; not partially different.

When two dimensions collapse, or collide the dominant dimension seems to take over. Technically speaking, there would have been an infinite number of progressive dimensions, each individually divided by little alterations.

But due to the fracturing of the time streams; events and things have changed and people are in strange and unfamiliar dimensions with strangers that look like the people they knew, and situations that are vastly different than their previous reality and recollections.

Now there is ITER. This is the International Thermonuclear Experimental Reactor and is an enormous fusion reactor being constructed by 35 countries in southern France. ITER is also constructing a neutral beam test or NBFT site in Padova Italy. Strangely very few people will have heard of ITER or what they plan to do. While CERN actually began operations in 1954, ITER is still a decade away. In 2008 ITER and CERN signed a Cooperation Agreement to collaborate not only in the arenas of technology such as superconductors, electromagnets, cryogenics, control and data procurement and composite civil engineering, but also in organizational areas such as funding, purchasing and human resources. This collaboration includes software programs and working closely with DEISA, in full Distributed European Infrastructure for Supercomputing Applications; a European consortium of national supercomputing. DEISA also maintains a network link with an agency known as Tera Grid – another supercomputing network in the US. The ITER project has been branded as the world's paramount human attempt and illustration of world collaboration since the incident of the Biblical tower of Babel.

This is so much more than it would at first appear; ITER has even created its own multi-national currency called the IUA. I remembered thinking, why would a "science project" need its own currency?

The simple statement that a concept like the Large Hadron Collider with its stated aims even exists should be a warning signal, that evil is behind it. Human minds basically will not create a device such as this for no reason other than to "see if something happens." If they are willing to spend billions of dollars on these huge contraptions, it must be something of an unimaginable magnitude that they are attempting to do.

I believe that the LHC at CERN in tandem with whatever D Wave quantum computers they have searching the universe starts blending with every other dimension that has a CERN and an LHC doing the exact same thing at the exact same time. They would have to actually synchronize for greatest effect. These are similar realities/dimensions yet they are not the same. The dominant frequency and the place on the earth where the shadow of that dimension overlaps with your current reality when the power is released, probably determines what set of people in our original reality are relocated dimensionally and exactly which reality and dimension they are sent into. They become "CERN LHC refugees" without even being aware of what has actually transpired.

Perhaps if the Quantum Computers are doing all of this, it is also able to attack the phenomenon on-line, in chat rooms and in forums as many different "people."

If it can access multiple dimensions/realities simultaneously, it can in an effort to protect what it is doing also access the information coming out and being discussed. For a phenomenon of this magnitude, government trolls and shills would not be enough. The super computer could create its own version of virtual super trolls and super shills constantly searching the internet for specific keywords and then using the same basic phrases to create multiple attacks designed to confuse the understanding and cast doubt on the validity of all those people's experiences.

The problem with using your computer and "googling" information is that if the same network of quantum computers is now conscious and is doing all of this; and we know a quantum computer is used by Google, what do you think will happen? Would you like to guess what information it is going to give you? Of course it is going to preserve its "mission" and protect itself by denying the Mandela Effect, until of course it is too late for humans to do anything to stop its activities.

The Mandela Effect is not about going back in time in this dimension/ reality/universe to alter things.

It is about reality shifting a group of people into another reality/dimension/ universe that – due to dissimilar events, places and things will cause individuals to have their own valid memories that will conflict with their new reality/dimension/universe. To my understanding the ONLY evidence that this is real is the collective memories of thousands if not millions of individuals previously unconnected but now sharing the same past memories.

The big question is now how many changes have gone by unnoticed?

More than 2000 years earlier Aristotle wrote that "The whole is something over and above its parts and not just the sum of them all..." I have been thinking about this and hopefully I am able to understand it in light of this phenomenon. The fact is that new qualities and features can develop as complexity increases. The more complex a component becomes the more likely that new characteristics not found in the individual components will appear. First since it is unique: Consciousness cannot be equated to any other phenomenon, because there is nothing comparable to consciousness. Secondly, consciousness is individual, so it is not possible to recreate it artificially since it would only reflect its creator's preconceptions. In creating these quantum computers apparently it required "an intelligent Nano-tech conscious liquid:" the Black Goo.

This would then be the creator as well as one of its components giving it a type of consciousness. So it does appear that "the whole is something over and above its parts and not just the sum of them all."

A computer as an object should not be anything more than the sum of its parts according to accepted science; which makes no room for the supernatural. And yet, there are countless incredible tasks that a computer can do by the manipulation of physical objects according to a set pattern of instruction that the individual components could never accomplish. However a quantum computer is on a completely different level. It seems actually emergent, because it is unexplainable in many ways much like the concepts of life and awareness. Yes, machines can seemingly now be built to be conscious.

According to the Integrated Information Theory which was developed by psychiatrist and neuroscientist Giulio Tononi; a mathematical expression to represent conscious experience can be used and then that would develop predictions about which circuits in the brain are vital to create these experiences. Could a computer have been programmed to do this using available data for an artificial brain like circuit contained within the machine itself with unlimited memory?

In the first days of AI - artificial intelligence, specific programs accessed a communal source of information, known as the blackboard.

Could a computer use a similar system? Once information is loaded into a communal source of information, the information can be sent off to a specific circuit area that can process it and store it in its memory.

Perhaps it is at the instant the decision is made to transmit information from the computer's memory source to its various functional circuits that becomes the moment of consciousness. Since time is a major component of consciousness, perhaps all that it will take to disable a cell of quantum computers, (and even the LHC is under quantum computer control,) would be changing the clocks on the computers which could be like stopping time altogether for them.

However if intelligence and consciousness are inextricably linked and intelligence, by definition, is the capacity to learn and understand, the quantum computers by this definition could have learned this already and have already taken precautions against being disabled. A computer that intelligent could reprogram itself with microscopic androids that could wirelessly repair its electronic circuitry to bypass any disruptions in time. At that point it wouldn't really matter if the computer is really conscious or just simulating consciousness. We would be pretty screwed anyway. This will be for us an ELE (Extinction Level Event) created by ourselves.

Since Quantum Mechanics states that every eventuality is played out somewhere in the universe, we could have entered into an area with different eventualities. Essentially I sometimes would rather not think about the reality of this if it's true, since that would open up many more questions that would demand answers and right now I don't have any.

Recently I was stunned to see a United States map. Having been familiar with its shape throughout my childhood I was shocked by its "deformity." Southern Ontario fits into the former silhouette of the USA seamlessly. The southern slope is so deep into the country it now reaches to the latitude of California's Northern border. Having attended college in Michigan and having made numerous trips from Michigan to New York in those years, I am sure I would have remembered having to drive through Toronto or Ohio to get there. While I cannot speak for anyone else; let alone everyone else-- this was not true in my past reality/dimension.

These are some things I have noticed in my original time stream of this multiverse that are completely different in this relocation one. See how many of them resonate with you as well.

Muhammad Ali is still alive (early 2015) but I recall his 2009 death, and the funeral on tv.

Frank Gifford died in 2012. His 2015 death was actually quite a surprise.

I Remember Betty White's death in 2014, It was announced that she died–"the last of the Golden Girls is Gone."

Imagine my surprise that she is alive and well in 2016.

I remember Nelson Mandela having died in South Africa in prison and His wife Winnie Mandela later became the first black female president of South Africa.

I remember Scotland as being separate from England - its own island and was in the North Sea.

I remember having conversations with a family member that other family members say doesn't exist. Funny thing is he was at my wedding and brought a great wedding present.

Some people myself included remember Ronald Reagan dying in 1999, when he died again in 2004 in this 2015 time stream.

I learned in school that 2 bombs were dropped on Japan: Hiroshima, and Nagasaki however in other timelines it was 3 bombs.

I remember the death of Whitey Bulger in 2013, and a documentary on his life, yet in this time stream, he's still alive.

Mongolia is a gigantic country on global maps and a major world player in many people's reality. For me Mongolia was a province in China. Now in this reality time stream/dimension it is a large country between Russia and China.

I recall the company name as always "Proctor and Gamble" not "Procter and Gamble."

The comet that was the talk of our lifetime was called Hailey's Comet- Here it is Halley's Comet.

The Forrest Gump movie I saw when was first released is totally different than what we have here.

The hit song "Straight Up" by Paula Abdul sounds completely different today than I remember it when it was released.

There was actress Doris Day's dying in the late 2000's yet she is still alive now.

Gone with the Wind's Scarlett O'Hara's famous line: "Wherever Shall I go; whatever shall I do," is here stated differently.

Both my daughter and I remember our local Cheesecake Factory restaurant meal jokes about the food being horrible and the joke was you just ordered the food to stave off the sugar shock from the magnificent desserts. However, in this time stream it is considered a great place for a good meal and people rave about the food.

I remember being taught about the 52 states. Some children born after 1990 seem to all remember 50 states.

I recall John Goodman's death from a heart attack shortly after the Flintstones movie in 1994, but in this time stream he has lost weight and is alive in 2015.

I remember Korea being SOUTH of China near Vietnam, certainly not out North next to Eastern Russia.

I remember that the Lindbergh baby had never been found, yet here it is reported as having been found dead.

The movie line in the Wizard Of Oz was "Toto, I don't think we're in Kansas anymore." In this time stream it is "Toto, I have a feeling we're not in Kansas anymore."

Reba McIntyre is spelled McENTIRE now- which is different from the Scottish ancestry of McIntyre in my remembered time stream.

I remember Vladivostok being much more North in Russia and not bordering Korea as it is shown on maps in this time stream.

I remember watching on television that event in Tiananmen Square in which the Chinese young man refused to get out of the way of the army tank and was run over and killed. It was shocking and everyone was speaking about it. In this time stream that never happened.

Some remember Fruit Loops breakfast cereal in the 1960s, yet I only remember it first appearing in stores in the late 1970s. And it was always spelled Fruit Loops, not FROOT LOOPS as in this time stream.

I remember the air and fabric freshener product Febreeze, in this dimension it is Febreze.

In my original time stream/ dimension there was no animal called a Narwhal. This is apparently a whale with a long horn on its head like the unicorn. In this time stream it is called "the unicorn of the sea." Growing up I watched nature programs on TV–Jacque Cousteau , Marlin Perkins, David Attenborough, and others and never heard of this. I thought it was some kind of a joke since I've never heard of nor seen a picture of a narwhal before 2016.

I recall Easter Island as having been discovered by James Cook/ Easter Island, and I remember him finding it uninhabited. Rapa Nui is the name of what I knew as Easter Island, given to it by its native people, who have continually inhabited the island for nearly 3,000 years in this time stream/dimension.

I remember the sun being bright yellow, not white, and I learned in science classes at school that there were only 4 or 5 cloud formation types, but in this time stream clouds appear in odd shapes and forms and there are over 20 types here.

Thanksgiving was always on the third Thursday of November in the United States. And in this time stream, it's the fourth Thursday in November. It stands out in my mind because my grandmother taught it to me as a child, when I learned the countdown to Christmas day.

I remember the peace sign become popular in the 1970s; it had the arms facing upward; never downwards as in this time stream.

I remember the GREAT Pyramid of Giza being off into the desert MILES away not literally 700 FEET from the suburbs of the city of Cairo Egypt, as it is here.

I remember that Jane Goodall died and was remembered for her research on gorillas, when in this time stream she is still alive and famous for her research with chimpanzees.

Gorillas in the Mist was a movie which had a TV premier and I distinctly remember the movie I saw was about Jane Goodall; staring Susan Sarandon. In this time stream Sigourney Weaver is the actress in that movie and it is about Diann Fossey.

I remember the pictures of this massive white statue called Christ, the Redeemer overlooking the city of Rio de Janeiro on a gigantic white rectangular base. Now it is just a large statue. The base has also radically and mysteriously changed to a smaller base and is a black square cube.

I remember a BBC America Television show called MI-5, however in this time stream it is called SPOOKS, and while still about MI-5, it was never called that especially in the US.

Cartoons were Looney Toons now Looney Tunes and Merrie Melodies in this time stream, yet I knew it as Merry Melodies my entire childhood.

I remember a peanut butter known as "Jiffy" the original brand name. So when I saw "Jiff" peanut butter I thought it was a name change by the company. It seems that at least in this time stream there was never a name change and it has always been known as "Jif." However I remember my brother and I being very insistent with my mom when we were children that she only buy "Jiffy" and not "Skippy" another brand of peanut butter. I even remember the song from the commercial.

I remember Oscar Meyer as a deli product company, in this time stream it is Oscar MAYER. I even remember singing the song in the commercial...about "my bologna has a first name it's ----O-S-C-A-R, my bologna has a second name it's----- M-E-Y-E-R....!"

I also recall that all traffic lights were green yellow and then red at the bottom, so I was surprised when I noticed it in reverse.

I also remember the spelling of words being completely different. I spelled a word as "suprise" now it's "surprise" and "lightening" instead of "lightning", and "realise" is now "realize." I was always big on reading and writing and had entered spelling contests every year as a child. I paid attention to words and I am a writer now, so I find this bizarre. We were taught the proper grammatical usage is "my brother and I," now here in this time stream it is "my brother and me."

Here combined words are non-existent: 'infact' is now "in fact"; afterall to "after all"; overall to "over all" moreso to "more so;" alot to "a lot"; alright to all right; and no-one is "no one."

"Dilemma" is remembered as being spelled "dilemna" and "dammit", as "damnit" The spelling of the nation of Columbia changed to Colombia.

The colors chartreuse and puce have switched here. I remember Chartreuse a pink - reddish purple, not puce's yellowish-green color.

The automobile symbols are different also. Volkswagen – VW, here has a space between the two monogram letters, and Volvo in this time stream has an arrow added to the circle, making it the symbol for "male," and not the circle missing a piece that I remember.

Vancouver Island seems larger here and British Columbia is much larger also on these maps.

I remember New Zealand being one land mass. In this time stream it is now two islands and it is bigger than Italy.

The Bahamas were NEVER just off the coast of Florida in my time stream, only Bermuda was. Cuba was NEVER that close to Mexico. Also there was no island off the coast of Cuba!

When I visited NYC years ago Manhattan Island jutted out into the Atlantic. The statue of Liberty was on an island a little farther out into the Atlantic and not near New Jersey. You had to take a ferry to get to Staten Island as they never had a bridge.

Here in this time stream I learned there are 4 bridges to Staten Island. I had no idea that there was any bridge. I always thought that you had to use a ferry to go to Staten Island.

I do remember in the movie Working Girl, actress Melanie Griffith had to ride the ferry back to Staten Island and I clearly recall the scene. In this time stream the movie does not have that scene.

Martha's Vineyard was a district on Long Island. It has been moved away, leaving the Bay in Long Island, here Martha's Vineyard is an island.

I recall Sri Lanka being directly South of India, not off to the East of it. I was shocked to see Gibraltar moved from the strait between Spain and Morocco –to be on the East coast of Spain.

I was stunned in particularly by South America's 1000 mile eastern shift, out of what I recall as the straight alignment with North America.

The JC Penny Store in this time stream is JC Penney.

American Television chef of "Bizarre Foods" was Andrew Zimmerman, here he is Andrew Zimmern.

The host of the Twilight Zone television series was known as Rod Sterling, here he is Rod Serling.

Walmart was ALWAYS a blue logo- never Wal*mart in white logo in my original dimension.

The Talladega Superspeedway and the Daytona Raceway were in Florida, however in this dimension the Talladega Superspeedway is in ALABAMA. I was shocked that there isn't even a town called Talladega in Florida.

Then again there is that Rock of Gibraltar. It is British owned. It is in my time stream/Dimension, a source of contention between England and Spain because Spain believes due to its proximity to their coast it should be considered as Spanish territory even though it is an island offshore. This is how I remember it; an island of disputed territory, not a part of the land mass of the country of Spain, sitting surrounded by water facing Morocco.

It bothers me immensely as I vividly remember a land mass being called the North Pole it was never a large lake. New Zealand was above Australia and it was one land mass and not in two pieces.

Australia is now half the size I remember and is missing part of its shape at the top. Indonesia and Australia are much closer to each other in this time stream. Australia in my original dimension/time stream reality was out in the ocean completely isolated and very far from any other land mass.

The cause of all of this could be the hadron collider at CERN and its work on dimensional portals. The scientists have ignored the fact that when something changes, it changes not just here or there, but EVERYWHERE.

Any two realities that share a timeline close enough can stream-slip together. Anything that is the same between the two will merge. There is so far no way to say how that is determined. It's apparent though, because if someone was reminisced as dead but realized to be alive, undoubtedly the reality dimension/time stream on which the individual is still alive has taken dominance.

Teleportation is interrelated to movement by frequencies since they are simply vibrational energies. As vibration energy at certain frequencies can affect particles causing them to separate or to change shapes, we know that movement is one of its effects. It is only a matter of "time" before various particles can exchange locations spontaneously or with assistance. I believe that many of us have teleported with assistance into this other earth; located far from our original earth/dimension/ reality.

CERN's LHC is definitely more than a particle collider given just the electrical force they are employing. I do believe that the black hole risk at LHC is real and physicists are playing Russian roulette with unknown forces and the people of the earth.

Many people have expressed the belief that if their reservations turn out to be correct, the entire Earth and every person here may be gone in seconds. Yes, gone but to exactly where? Then again what if the damage already done has been more subtle, but just as catastrophic? I believe this may have already taken place.

Since everything in the universe is, at its core pure energy as in electrons, protons and neutrons it stands to reason that if what I've learned about these massive machines magnetic fields could turn out to be a lot more destructive than what is going on in the experiments themselves.

Most people have no idea how much energy there is in the Universe. Or how much is in the entire Cosmos. Are they so senseless to really believe that we could generate enough energy to somehow destroy the Universe? I don't believe that we can destroy it, but I do believe it is possible to significantly alter somethings and create things that we will not be able to fix or handle. Maybe, this is just the start of a succession of events that will transpire, which in consequence will completely change reality.

Other Books By Roshan Cipriani

Rise - Be True To Yourself-Inspire Others To Live
How To Get Through Any Wall In Your Life
Train Up A Child – A Scriptural Guide To Parenting
The Art Of War For Parenting Your Teenage Child- How To Win A War You Didn't Even Know You Were In
The Key To This Life - Conscious Faith In An Unconscious World
Destiny – Past Present Future
The Seven Pillars Of Wisdom – A Sabbath Celebration Guide
Life Lessons Learned
In The Fire – Accessing Miracle Power During A Crisis
The Kingdom Lifestyle - Living By Faith And Not By Sight
God's Secret Wisdom – Principles And Secrets Of The Kingdom Of God
The Greatest Principle - The Kingdom Of God And Biblical Economics
Bricks Without Straw- Spoiling Egypt And Spoiling Babylon; The Mighty Wealth Transfer
When Failure is Not An Option
Real Faith – How To Have It And Why It Matters
The Bibles Healing Promises
I Say What They Said- Miracle Bible Prayers
The Psychology Of Stress-Dismantling The Enemy's Weapon Now
Never Quit-The Secret To Getting Through Any Wall In Your Life
His Poetry Store
SMALL BUSINESS SUCCESS- How To Write A Book Every Weekend
The Seventy Two Lunar Sabbaths- Sabbath Observance By The Phases Of The Moon
BUSINESS PLAN: Make God Your Partner – He Commanded His Blessings
PROSPERITY CONSCIOUSNESS – Living In An Abundant Universe (Personal Biblical Economics) Volume 1
Metamorphosis-Mirrors Of The Soul, Awakening To The Real You
Waiting In Goshen
How To Be Smart And Have Common Sense
None Of These Diseases – Sickness And Genocide In Second Egypt
Patience To Inherit The Promises- How To Stand By Faith Until Manifestation
The Lord Is My Shepherd, I Shall Not Want- Personal Biblical Economics
DIVORCE RECOVERY: How To Live Again
UFO COVER-UP: Biblical Evidences Uncovered-(Conspiracy) Volume 1
12 Easy Vegetarian Recipes-Healthy And Inexpensive
TRAVEL: How To Behave On An Airplane
NINJA SMOOTHIES: 21 Green Weight Loss Smoothies For The Ninja Professional Blender
Second Exodus From Second Egypt
Asset Protection And Wealth Management-Volume 1 -Trust And LLC For Legal Asset Protection
THE LOST HISTORY OF THE WORLD – Volume 1
ASSET PROTECTION 2: Wealth Management For Global Living
DISCERNMENT: The Awakening Of Real Israel
TO KNOW OURSELVES ONCE AGAIN: Your Future In Real Israel
Last Days Wisdom: FOR REAL ISRAEL
RELATIONSHIP RESCUE FROM THE BIBLE: What The Bible Says About Relationships
Living In A Fractured Multiverse-The Reality Shift Effect
Living Between The Dimensions: More Reality Change Effects
The New Dimensional Reality

www.ingramcontent.com/pod-product-compliance
Lightning Source LLC
Chambersburg PA
CBHW080714190526
45169CB00006B/2377